BRIDGING THE LEVEN

By
Gordon Burns

First Published February 2009
Second Edition August 2009

Published by the author

The author / publisher has made every effort

to trace the provenance of the photographs in this book.

No breach of copyright, if in existence at the

time of publication, was intended.

© Copyright 2009 – Gordon Burns

Prelude

Crossing a bridge is like getting on a plane or boarding a ship, you do not know what you are going to find until you arrive at the other side.

My original interest came from the rebuilding of the Bonhill Bridge in 1987.

Oh, ye 'll tak the high road,

and I 'll tak the low road;

And I 'll be in Scotland afore ye:

But me and my true love

will never meet again:

On the bonnie, bonnie banks

o ' Loch Lomond.

The River Leven

From Loch Lomond............

............to the River Clyde.

There are ten bridges which cross the River Leven in the relatively short five mile distance from the southern end of Loch Lomond at Balloch, to entering the River Clyde at Dumbarton.

They are as follows :-

Balloch Bridge – vehicular and pedestrian.

Renton Bridge – pedestrian only.

Lomond Road Bridge – vehicular and pedestrian.

Leven Bridge – Alexandria By-pass (A82) vehicular only.

Stirling (Stuckie) Bridge – originally a railway bridge, now pedestrian only.

Dalreoch Bridge – railway only.

Bonhill Bridge – vehicular and pedestrian.

Artizan Bridge – vehicular and pedestrian.

Dillichip (Black) Bridge – Originally a railway bridge, then pedestrian only, now merely a structure.

Dumbarton Bridge – vehicular and pedestrian.

Introduction.

The Vale of Leven has, over the centuries, been on the route between the west coast of Scotland to the north of the River Clyde serving the movement of people, livestock and transportation from there to the cities, towns and market places in the central belt of Scotland.

The River Leven is still navigable its full length, with the original towpath, now cycle path, on the west bank, and is tidal to Dalquhurn Point.

The name 'Leven' is derived from the Gaelic 'Leamhan' meaning Elm Bank, and the river is the second fastest flowing river in Scotland.

The River Leven is in geological terms referred to as a meandering river due to the sandy alluvial nature of the soil deposited during the last ice age when Loch Lomond was carved out by glaciers. This allowed the water to be filtered sufficiently giving rise to the textile industries which existed in the Vale of Leven area from before the Jacobite Rising and through the Industrial Revolution, lasting over 200 years, until the 1960's.

Methods of crossing the river have existed in various forms from fords, small ferries, narrow pedestrian bridges, to full scale dual carriageway road bridges.

However, this type of soil does not allow for good solid foundations for the building of bridges where heavy traffic is constantly pounding its way over them, and so "adjustments" to their construction have had to be made over time. From the earliest to the most recent, constant monitoring is vital.

Balloch Bridge

ORIGINAL BRIDGE, BALLOCH.
BUILT 1841.

The original bridge was built in 1841 by Sir James Colquhoun of Luss, a chain suspension construction, supported by stone towers and similar to the "Bawbee Brig" at Alexandria and Bonhill. It only had a capacity of under three tons, and a span of 280 feet.
This replaced the toll ferry installed around 1814, providing a safer crossing, to serve the ever increasing population arising from the expansion of the textile industry in the area.

SECOND BRIDGE, BALLOCH.
BUILT 1887.

A second bridge was opened in 1887 of steel girder construction on red sandstone piers, over a full length of five spans, 25 feet wide, with one pedestrian footpath and a two-way tarmacadam vehicular carriageway. It cost £5000, the main contractors being Messrs. Hanna, Donald and Wilson of Paisley, masonry sub-contractor was William Barlas of Dumbarton, and the engineers were Messrs. Crouch and Hogg of Glasgow.
In November 2003 additional upgrading work to the main structure was carried out at a cost of £1.25 million. The whole decking was removed leaving only the red sandstone supports.

THIRD BRIDGE, BALLOCH.
BUILT 2004.

The new steel decking now meets current load-bearing requirements, and extended in cantilever fashion to provide footpaths, cycle ways and viewpoints on each side. The steel and wire railings were built to a height of 1.4m (normally 1m) and angled inwards to try and prevent people jumping into the river.

The bridge was completed and opened on 3 September 2004, and traffic was diverted via the Lomond Road Bridge for the duration of the works.

NEW DECK SHOWING CANTILEVERED FOOTPATH

Lomond Road Bridge

LOMOND ROAD BRIDGE, BALLOCH.
BUILT 1935.

This bridge was opened in July 1935, located on the A811 road, designed to connect the Dumbarton to Tarbet road with the Dumbarton to Stirling road.

Work commenced in July 1931, but was stopped in November 1932 because of a government economy measure, resuming the following August.

It carries a 6.1m wide carriageway and two 3m wide pavements edged with cast iron balustrades. The structure has three spans consisting of two steel plate girders supporting a reinforced concrete slab deck.

A plaque commemorating the opening of the bridge is missing, although the fixings can still be seen on one of the stone parapets.

The bridge deck is founded on abutments and intermediate piers constructed of reinforced concrete, finished in ashlar facing block. The central suspended main span is 22m long and the two side spans are both 19.5m long.

Messrs. Blyth and Blyth were the civil engineers, earthworks were by Messrs. J.A. Paton and Sons, Alexandria, foundations by Messrs. Crowley and Russell, and Messrs. William Arrol were responsible for the steelwork.

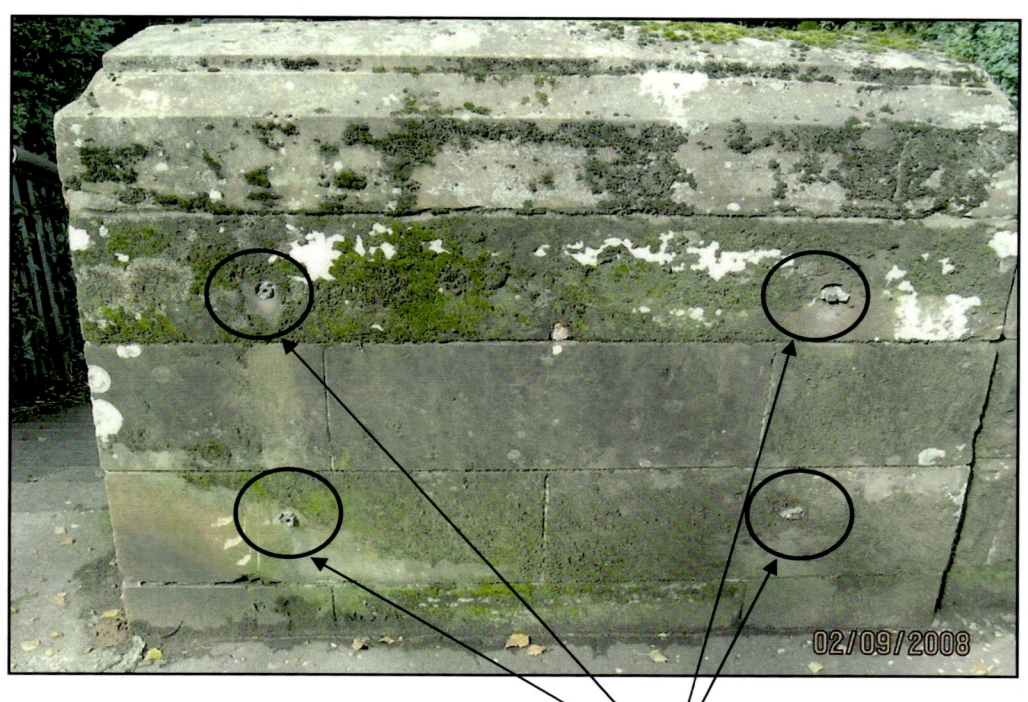

LOMOND ROAD BRIDGE, BALLOCH.

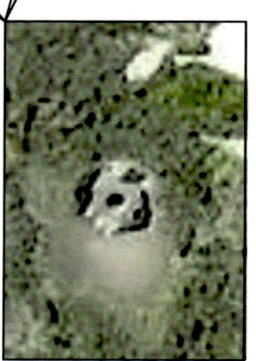

SITE OF MISSING PLAQUE

Stirling Bridge (Leven Viaduct)

SECOND STIRLING (STUCKIE) BRIDGE.
BUILT APP. 1877.

Sometimes called the Drymen Bridge, but perhaps known better locally as the Stuckie Bridge, a corruption of the word stirling to starling, (stuckie being the local word for a starling).

This bridge was constructed in approximately 1877, and was originally a railway bridge which carried a branch line from the main passenger and goods railway line running through the Vale of Leven ending at Balloch pier.

The line was owned by the Forth and Clyde Junction Railway Company, and ran on to Caldarvan, Buchlyvie and Stirling, but was closed to passengers in 1934 by LNER.

It has since been used as a pedestrian access providing a route between Alexandria at Rosshead over the Leven to Jamestown and Mill of Haldane.

The structure is a continuous three span twin steel truss supporting a reinforced concrete slab deck. The bridge deck is founded on stone abutments at each end, with the river span supported by two sets of steel piling or tubular supports which are set deep into the soil on each side of the river bank. The overall length of the bridge is 89m, the main span over the river being 38m.

This bridge superseded a wooden structure which was the first to carry the railway line opened on 20 May 1856, lasting only some twenty years before the steel one was built.

ORIGINAL STIRLING (STUCKIE) BRIDGE.
BUILT 1856.

Bonhill Bridge

ORIGINAL BONHILL BRIDGE. (BAWBEE BRIG)
BUILT 1836.

In 1836, the first pedestrian and vehicular crossing was made at this point on the River Leven.

A chain suspension bridge supported on stone piers, at the instigation of local landowner, Admiral Smollett, was erected to replace the chain link ferry and ford which had previously existed. It had a span of 40m, width of 4.5m and a load capacity of three tons. The landing steps for the ferry can be seen on the west bank of the river in the picture.

It was nicknamed the " Bawbee Brig " due to the ha'penny charge made for a person to cross it (doubled between the hours of 11pm and 5am).

BAWBEE BRIG TOLLS

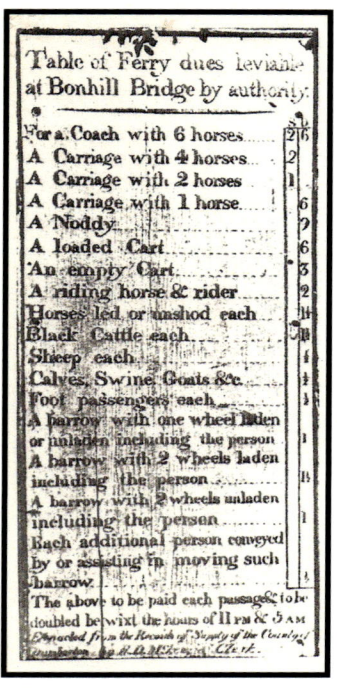

A Noddy	9 pennies
A coach with 6 horses	2 shillings and sixpence
A sheep	1 ha'penny each
A barrow with 2 wheels unladen including the person	1 penny

Tolls varied according to the type of traffic. (refer to list of charges above).
(A Noddy is a type of two-wheeled cab, having a door at the back, and a seat for the driver in front, drawn by a single pony.)

Before 1707, the Union of the Parliaments, a bawbee was worth a Scots sixpence, but adopted the ha'penny English value after the Union.

Prior to 1833, roads and bridges were the responsibility of local landowners. To recover expenses, tolls were charged to the public, and collected by Admiral Smollett causing considerable discontent. A committee was formed in 1848 to protest, but three years later, by which time the bridge should have paid for itself, the Road Trustees had failed to take conciliatory action. Riots prevailed.

Following the Roads and Bridges Act of 1878, the Bawbee Brig passed into public ownership, but the Road Trustees were obliged to continue toll charges, and was not settled until 1895, when the bridge was made free of tolls. (refer to 1895 poem The Auld Brig).

Written in May 1895 on freeing of bridges from tolls,
and published in the LENNOX HERALD.

The Auld Brig

Oh the Auld Brig, the Auld Brig,
Tho` noo ye`re bent an` grey,
Yet weel I min` in Auld Lang Syne,
Ye were sae fresh an` gay.

But noo ye`re auld an` thin an`bauld,
Ye`re hingin` sair agee,
For on yer heid like weicht o` lead,
Is the curse o` your bawbee.

Oh! the auld lairds, the auld lairds,
They`re gaun, the lairdies three,
O` a` they got, they`ve left us nought,
But this dear memorie.

Oh! the auld brig, the auld brig,
Ye`re jist aboot tae dee,
Yer sides are thrawn, yer cheeks are fa`in,
Ye`re shakin` like a tree.

An` noo, jist like the lairds, ye`ll work,
Nae mair iniquitie,
But meet yer fate, an` gang yer gait,
Without a broon bawbee.

A.S. MacBride.

Due to the poor state of repair and limited capacity of the Bawbee Brig, the Road Trustees arranged for a replacement, and on 2 July 1898 the " Whipple Arch " steel bridge was opened.

SECOND BONHILL BRIDGE. "WHIPPLE ARCH"
BUILT 1898.

It had a span of 50m and a width of 11.5m. It was supported on four red sandstone abutments and carried the usual utilities, as well as the tramlines for this new form of transport in the area.

Tram crossing the second Bonhill Bridge.

This bridge was sufficiently capable of coping with all the traffic between the original village of Bonhill and the growing town of Alexandria.

However, by 1962, following investigation, strengthening work was carried out. Remedial repairs failed to eradicate all its weaknesses, and the bridge was classified as unsuitable for vehicles over fourteen tons. Because of this, traffic lights were erected and phased to ensure there was no standing traffic on the bridge.

This, together with the expansion of Bonhill and redevelopment of Alexandria in the 1970's, another bridge was designed to meet modern criteria, aligning with re-routed road systems in each town.

To reflect the style of its predecessor, this was built in the form of a single-span, steel-tied through-arch, mirroring the only other one of its type in Scotland at Bonar Bridge in Sutherland, built in 1973.

BONAR BRIDGE, SUTHERLAND.
BUILT 1973.

Third bridge under construction

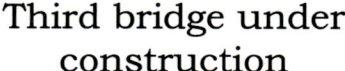

Third bridge under construction

Bonhill bridges, old & new side by side

The bridge cost £700,000, part of the total cost of £4.7m of the revised road and bridge layout. It is made of 284 tonnes of steel, with a span of 60m, width 14.5m, a reinforced concrete deck on steel plate girders, tie-beams and hangers, and supported on reinforced concrete abutments.

THIRD BONHILL BRIDGE. "RAINBOW BRIDGE".
BUILT 1987.

The steelwork was by Messrs. William Arrol. This turned out to be the last bridge the company built, while the 1898 "Whipple Arch" bridge was the first bridge they built!
The bridge was nicknamed the "Rainbow Bridge" and the first car drove over it on 28 July 1987.

On the Alexandria side of the bridge, three pyramids have been erected using some of the red sandstone from the 1898 bridge piers, and the plaque commemorating the opening of that bridge is incorporated in one of them. Behind them stand the remains of the four gas lights which stood on the four piers.

Dillichip Bridge

DILLICHIP (BLACK) BRIDGE.
BUILT 1876.

Sometimes known as the "Black Bridge", due to the fact that all the steelwork was painted black.
It was originally a railway bridge on the siding line off the main Dumbarton to Balloch line, using "pugs" to shunt wagons to and from the junction to the Dillichip Works on the east bank.
There was even a pug called "Dillichip"!

DILLICHIP BRIDGE.
THE "DILLICHIP PUG"

After the works closed due to a fire in the textiles store and the Army ceased to use it, the bridge was used as a useful vehicle for slinging gas and water pipes across the river.

When the railway line was removed, a footpath was created providing a pedestrian link between old and new Bonhill, the Vale of Leven Academy and Renton.

Built in 1876, the steel structure has three truss spans.

The main truss over the river is 28m long and is supported on two sets of twin piers built of masonry block, one set founded on rock below the river bed and the other on the river bank. The end trusses, 15m and 17m long, span between the piers and masonry abutments on each river bank. The total length is 61.5m.

It was closed to the public in December 2004 after a hole was discovered in the structure, then completely stripped, leaving only the girders.

DILLICHIP BRIDGE.
REMAINS OF STRUCTURE.

Flooding during December 2006 caused the east embankment to slip, jeopardising the bridge supports.

Renton Bridge

RENTON BRIDGE.
BUILT 1952.

This bridge was built in 1952 to provide pedestrian access only from the newly formed Vale of Leven Industrial Estate on the east side of the River Leven to the workforce on the west side, mainly from Renton. Vehicular access to the Estate is from the A813 road between Bellsmyre in Dumbarton and Bonhill.
Companies like Polaroid, Westclox, Burroughs, Turnkey and many more now provided work to replace the vanished textile industries.

RENTON BRIDGE.
BUILT 1952.

The three span right angled steel structure consists of a continuous truss girder, supported on reinforced concrete twin column piers. Access is by an approach deck and single flight stairway to the west, and a double flight stairway to the east side.
The span over the river is 45m long, and the total length is 95m.

Leven Bridge

VALE OF LEVEN BY-PASS BRIDGE. (A82)
BUILT 1970.

This bridge was built in 1970, as part of the re-routing of the main trunk road, the A82, serving the west coast of Scotland, extended from Barloan roundabout in Dumbarton to the Stoneymollan roundabout at Balloch, thus by-passing Dumbarton, Renton, Alexandria and Balloch.

It is constructed of a continuous five span steel plate girder with a reinforced concrete slab deck. The longest span over the river is 24.4m, and the total length is 102.6m. The bridge is supported on reinforced concrete abutments and piers.

A large steel water mains pipe (600mm diameter) is also supported by the bridge, and can be seen between the carriageways. It runs below ground level on each side of the bridge.

VALE OF LEVEN BY-PASS BRIDGE. (A82)
(I-BEAM CONSTRUCTION)

VALE OF LEVEN BY-PASS BRIDGE. (A82)
(SHOWING 600mm DIA. STEEL WATER PIPE.)

Dalreoch Railway Bridge

DALREOCH RAILWAY BRIDGE.
BUILT 1850.

Network Rail is now responsible for this bridge.
It was opened on 15 July 1850 to carry the railway line owned by the Caledonian and Dunbartonshire Junction Railway Company from Bowling to Balloch pier, and also the line to Helensburgh which opened on 31 May 1858.
It is built on red sandstone piers supporting four steel plate girders and reinforced concrete slabs carrying the tracks. The total span over the river is 91.4m.
Prior to the railway age, goods like timber, slate and textiles were conveyed by gabbart (a sail boat) on the Leven or by carts on the roads. A second proposal, (the first had been in the previous century), for canalising the River Leven put forward in 1840 stood little chance of succeeding with early railway promotion.

A short time after, the Glasgow, Paisley and Greenock Railway Company envisaged having a ferry link between Langbank and Dumbarton, and then a railway between Dumbarton
and Balloch
However landowners Colquhoun of Luss and Smollett of Bonhill wanted a line with more local control, and after proposals for a Glasgow to Balloch line ran into difficulties, a line from Bowling to Balloch won approval and opened on 15 July 1850.

Tourism was expanding rapidly at this time.

The bridge also witnessed the transition from steam trains to electrification in the 1960's.

DALREOCH RAILWAY BRIDGE.
(MAINTENANCE PLATFORMS.)

Artizan Bridge Dumbarton

ARTIZAN BRIDGE, DUMBARTON.
BUILT 1974.

This bridge was built as part of the redevelopment of Dumbarton town centre, and the construction of a new road through the town to accommodate the increased volume of traffic emanating from the expanding dormitory towns of Helensburgh and Cardross.
A plaque can be seen at the College Way pedestrian underpass commemorating the completion of the whole project, which was designed by two Dumbarton Academy former pupils.

It is built on the box girder principle, which is the same as the Erskine Bridge built in 1971, and the Westgate Bridge in Melbourne, Australia, built 1968-1978.

ARTIZAN BRIDGE, DUMBARTON.
BUILT 1974.
(BOX GIRDER CONSTRUCTION.)

ERSKINE BRIDGE.
BUILT 1971.
(BOX GIRDER CONSTRUCTION.)

WESTGATE BRIDGE.
MELBOURNE, AUSTRALIA.
BUILT 1968-1978
(BOX GIRDER CONSTRUCTION.)

ERSKINE BRIDGE.
(BOX GIRDER CONSTRUCTION.)

The box girders are supported by four reinforced concrete piers, two on land and two in the river, finished in tarmac with two wide concrete slab pavements, steel railings and expansion joints at each end.
Work began in 1971 at a total cost of £1.25m, the bridge itself costing £300,000.
Following the Napoleonic War, Dumbarton's Prison was built in 1824. It was closed in 1883, and demolished in 1973 to make way for the new road linking with the bridge. The stone main entrance portico and east wall with two original cell windows are all that remains of it adjacent to Glasgow Road.

DUMBARTON PRISON REMAINS.
BUILT 1824, DEMOLISHED 1973.

EAST WALL CELL WINDOWS.

MAIN ENTRANCE PORTICO.

The bridge was opened on 15 November 1974. Messrs. Babtie, Shaw and Morton were the consulting civil and structural engineers, consultant architects were Messrs. Garner, Preston and Strebel, and the contractor was Messrs. Miller Construction.

Dumbarton Bridge

DUMBARTON BRIDGE.
BUILT 1765.

This is the oldest bridge crossing the River Leven. It was built in 1765 of stone construction, partly at the instigation of the Duke of Argyll, who regularly travelled the route.

A ford existed at this site prior to the bridge being built. It was widened in 1884, reconstructed in 1934, and again in 2005.

The original bridge comprises five arches with three spans of 62 feet, two spans of 42 feet, a width of 20 feet, and an overall length of 360 feet. John Brown of Dumbarton was the architect.

One of the piers sunk into the river bed, and a report by John Smeaton, the Yorkshire born engineer, dated 31 May 1768, suggested a new pier be built on top, and be as light as possible to prevent further movement. He was regarded as the father of civil engineering in Britain, and is also responsible for the construction of the Forth and Clyde Canal.

The bridge is a category "B" listed building, but in its original condition would have been category "A".

In 1884, the bridge was widened by the addition of 8'6" wide footpaths on each side constructed of lattice steel, supported by cast iron cantilevers at 40 feet centres built into the stonework.

In 1934, major improvements were carried out by way of reinforced concrete to strengthen parapets, pavements, roadway and pier walls. The bridge also carries the usual utilities.

DUMBARTON BRIDGE.
SHOWING CONCRETE BALUSTRADES, CANTILEVERED FOOTPATH & PIER REPAIRS.

In 1999, £180,000 was spent repairing masonry, parapets, and the cantilevered concrete slab pavements.

From February 2001 until September 2005, a £1.75m makeover took place. The stone balustrades had, unfortunately, to be replaced with machined concrete ones.

Summary

So there it is, ten bridges across the River Leven. From roughly the middle of the 18th century until the present day, these bridges have served the communities between Loch Lomond and the River Clyde, expanding and changing to meet the requirements of the period.

They vary in size, design and function according to their location on the river. They have been altered in size and load bearing capacity as the communications infrastructure in the area has expanded to meet the needs of industry, population growth, increased volume and weight of road vehicles, greater movement of people and goods generally throughout the country. Constant monitoring is required to keep all of the above accessible to the public, and keep abreast with modern day living.

Long gone are the days when it only cost nine old pennies for a noddy, or a ha'penny for a person to cross the "Bawbee Brig" at Bonhill. Traffic now speeds through the Vale of Leven along the A82 trunk road from Glasgow to Fort William and beyond, probably hardly noticing the bridge at Dalmoak, allowing them passage over the River Leven.

Or any of the other nine bridges we have mentioned!!

The other significant structure on the river is the weir built at Balloch by Scottish Water between the Lomond Road Bridge and the Stuckie Bridge, to control the level of water contained in Loch Lomond.
No traffic crosses here.

Gordon Burns